MENU-MATH

Bacon and Eggs	$3.99	Lemonade	$ .77
Banana Split	2.16	Milk	.95
Cake	1.52	Milk Shake	1.86
Chicken	4.74	Onion Rings	1.63
Chili	1.89	Pancakes	2.55
Cola	.74	Pizza	1.17
French Fries	.90	Root Beer	.92
Hamburger	1.75	Sandwich	4.43
Hot Chocolate	.94	Shrimp	4.10
Hot Dog	1.10	Spaghetti	1.69
Ice Cream Cone	.88	Sundae	1.66
Juice	.75	Taco	.98

A Remedia Publication

MENU MATH FOR BEGINNERS

REM 111A

A TEACHING RESOURCE FROM...

REMEDIA PUBLICATIONS

BLACKLINE MASTERS

To find Remedia products in a store near you, visit:
http://www.rempub.com/stores

REMEDIA PUBLICATIONS, INC.
15887 N. 76TH STREET • SUITE 120 • SCOTTSDALE, AZ • 85260

207

Certificate
of Achievement

________________________________ has, on this

day, the ____________ of ________________________,

in the year ______________, successfully completed the

Menu-Math Book.

Congratulations!

______________________ ______________________

Teacher School

LOCATING PRICES ON THE MENU

Use the menu to find the price of each item.

1.  $ ____________

2.  $ ____________

3.  $ ____________

4.  $ ____________

5.  $ ____________

6.  $ ____________

7.  $ ____________

8.  $ ____________

9.  $ ____________

10.  $ ____________

11.  $ ____________

12.  $ ____________

13.  $ ____________

14.  $ ____________

15.  $ ____________

16.  $ ____________

LOCATING PRICES ON THE MENU

Use the menu to find the price of each item.

1.	Banana Split	$ ________	11.	Pizza	$ ________
2.	Cola	$ ________	12.	Juice	$ ________
3.	Ice Cream Cone	$ ________	13.	Milk	$ ________
4.	Onion Rings	$ ________	14.	Hot Dog	$ ________
5.	Hamburger	$ ________	15.	Taco	$ ________
6.	Sandwich	$ ________	16.	Cake	$ ________
7.	Chicken	$ ________	17.	Chili	$ ________
8.	Root Beer	$ ________	18.	Spaghetti	$ ________
9.	Sundae	$ ________	19.	Pancakes	$ ________
10.	Shrimp	$ ________	20.	Lemonade	$ ________

ADDITION - 2 items

No regrouping

Use the prices from the menu.
Find the total cost.

1.  $ ________

    $ ________

   Total $ ________

2.  $ ________

    $ ________

   Total $ ________

3.  $ ________

    $ ________

   Total $ ________

4.  $ ________

    $ ________

   Total $ ________

5.  $ ________

    $ ________

   Total $ ________

6.  $ ________

    $ ________

   Total $ ________

7.  $ ________

    $ ________

   Total $ ________

8.  $ ________

    $ ________

   Total $ ________

ADDITION - 2 items

Use prices from the menu.
Find the total cost.

1. 

$__________

2. 

$__________

3. 

$__________

4. 

$__________

5. 

$__________

6. 

$__________

7. 

$__________

8. 

$__________

9. 

$__________

10. 

$__________

ADDITION - 2 items

Use prices from the menu.
Find the total cost.

---

1. Two  cost $ ____________

2. Two  cost $ ____________

3. Two  cost $ ____________

4. Two  cost $ ____________

5. Two  cost $ ____________

6. Two  cost $ ____________

7. Two  cost $ ____________

8. Two  cost $ ____________

9. Two  cost $ ____________

10. Two  cost $ ____________

ADDITION - 2 items

Use prices from the menu.
Find the total cost.

1. Banana Split $ ______ Sundae $ ______ Total $ ______	5. Milk Shake $ ______ Spaghetti $ ______ Total $ ______
2. Pancakes $ ______ Juice $ ______ Total $ ______	6. Hamburger $ ______ Cola $ ______ Total $ ______
3. Shrimp $ ______ Sandwich $ ______ Total $ ______	7. Lemonade $ ______ Chicken $ ______ Total $ ______
4. Cake $ ______ Onion Rings $ ______ Total $ ______	8. Bacon and Eggs $ ______ Milk $ ______ Total $ ______

ADDITION - 2 items

Use prices from the menu.
Solve each problem.

---

1. One  and one  cost $ ____________

2. One  and one  cost $ ____________

3. One  and one  cost $ ____________

4. One  and one  cost $ ____________

5. One  and one  cost $ ____________

6. One  and one  cost $ ____________

7. One  and one  cost $ ____________

8. One  and one  cost $ ____________

9. One  and one  cost $ ____________

10. One  and one  cost $ ____________

ADDITION - 2 items

Use prices from the menu.
Solve each problem.

1. How much for this food?  

   $ ____________

2. Buy  and a  . How much in all?

   $ ____________

3. Bill buys a  and a  . What is the cost?

   $ ____________

4. How much for a  and a  ?

   $ ____________

5. Buy a  and a  . How much in all?

   $ ____________

6. Jeff has a  . Kim has  . What is the cost?

   $ ____________

7. How much for this food?  

   $ ____________

ADDITION - 2 items

Use prices from the menu.
Solve each problem.

1. Mike had a sandwich and milk.
   How much did he spend? $ ________

2. How much for pancakes and juice? $ ________

3. What do two hot dogs cost? $ ________

4. Sara buys french fries and a cola.
   What is the total cost? $ ________

5. How much for cake and root beer? $ ________

6. Susan bought a taco. Then she bought a sundae. How much did she spend? $ ________

7. How much for chili and lemonade? $ ________

ADDITION - 3 items

Use prices from the menu.
Find the total cost.

1.  $ ________

 $ ________

 $ ________

Total $ ________

2.  $ ________

 $ ________

 $ ________

Total $ ________

3.  $ ________

 $ ________

 $ ________

Total $ ________

4.  $ ________

 $ ________

 $ ________

Total $ ________

5.  $ ________

 $ ________

 $ ________

Total $ ________

6.  $ ________

 $ ________

 $ ________

Total $ ________

ADDITION - 3 items

Use prices from the menu.
Find the total cost.

1.

$ ____________

2.

$ ____________

3.

$ ____________

4.

$ ____________

5.

$ ____________

6.

$ ____________

7.

$ ____________

8.

$ ____________

ADDITION - 3 items

Use prices from the menu.
Solve each problem.

1. How much for this food?    $ ________

2. Buy a  , a  ,and a  .

   How much in all? $ ________

3. Ed bought  , a  , and a  .

   What is the total cost? $ ________

4. How much for  ,  , and an  ? $ ________

5. Sue bought a  , a  , and  .

   How much did she spend? $ ________

6. Buy  , a  , and  .

   What is the total cost? $ ________

ADDITION - 3 items

Use prices from the menu.
Solve each problem.

---

1. A man bought  ,  , and  .

   How much did that cost? $ ________

2. Steve bought  ,  , and  .

   How much did he pay? $ ________

3. If you buy  ,  , and  , how much will you owe? $ ________

4. Ann ordered  ,  , and  . What was the total cost? $ ________

5. How much for  ,  , and  ?

   $ ________

6. Jack ate a  ,  , and  . What was the total cost of this food? $ ________

ADDITION - 3 items

Use prices from the menu.
Solve each problem.

1.	Taco	$ ________
	Root Beer	$ ________
	Sundae	$ ________
	Total	$ ________
2.	Lemonade	$ ________
	Milk Shake	$ ________
	Milk	$ ________
	Total	$ ________
3.	Sandwich	$ ________
	French Fries	$ ________
	Cake	$ ________
	Total	$ ________
4.	Chicken	$ ________
	Onion Rings	$ ________
	Cola	$ ________
	Total	$ ________
5.	Pancakes	$ ________
	Juice	$ ________
	Hot Chocolate	$ ________
	Total	$ ________
6.	Hot Dog	$ ________
	Pizza	$ ________
	Banana Split	$ ________
	Total	$ ________

ADDITION - 2 and 3 items

Use prices from the menu.
Solve each problem.

1. Tracy bought a sundae. Then she bought a cola. How much did she spend? $ ________

2. How much for pancakes, juice, and hot chocolate? $ ________

3. Dave buys spaghetti and lemonade. What is the total cost? $ ________

4. Terry had a hamburger, french fries, and a milk shake. How much did he spend? $ ________

5. How much for pizza, root beer, and a sundae? $ ________

6. How much for bacon and eggs and hot chocolate? $ ________

ADDITION - 3 items

Use prices from the menu.
Solve each problem.

1. Pete had pizza, root beer, and a banana split. How much did he spend? $ ________

2. How much for bacon and eggs, pancakes, and juice? $ ________

3. What do a hamburger and two colas cost? $ ________

4. Mr. White buys a sandwich, a milk shake, and a sundae. What is the total cost? $ ________

5. How much for shrimp, onion rings, and cake? $ ________

6. Joe bought a hot dog and milk; then he bought a banana split. How much did he spend? $ ________

SUBTRACTION

No regrouping

Use prices from the menu.

Solve each problem.

Had	Bought	Had Left
1. $3.58		
2. $2.86		
3. $4.68	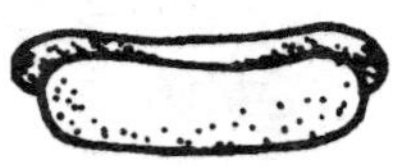	
4. $3.97		
5. $4.88		
6. $8.65		
7. $3.49		
8. $5.75		

SUBTRACTION
No regrouping.

Use prices from the menu.
Solve each problem.

1. Had $.99

Bought 

How much left?
$ ____________

2. Had $1.75

Bought 

How much left?
$ ____________

3. Had $.89 

Bought

How much left?
$ ____________

4. Had $.86 

Bought

How much left?
$ ____________

5. Had $1.96

Bought 

How much left?
$ ____________

6. Had $2.28

Bought 

How much left?
$ ____________

7. Had $5.64

Bought 

How much left?
$ ____________

8. Had $3.87

Bought 

How much left?
$ ____________

SUBTRACTION

Use prices from the menu.
Solve each problem.

You Want	You Have	How much more do you need?
	$ .90	
	$ .25	
	$ .54	
	$1.09	
	$ .46	
	$1.38	
	$ .53	

SUBTRACTION

Use prices from the menu.
Solve each problem.

Which Costs More?

1.  or  ______________________________

How much more? $ ___________

2.  or  ______________________________

How much more? $ ___________

3.  or  ______________________________

How much more? $ ___________

4.  or  ______________________________

How much more? $ ___________

5.  or  ______________________________

How much more? $ ___________

SUBTRACTION

Use prices from the menu.
Solve each problem.

1.  

$ ________ $ ________

How much more for

 ? $ ________

2.  

$ ________ $ ________

How much more for

 ? $ ________

3.  

$ ________ $ ________

How much more for

 ? $ ________

4.  

$ ________ $ ________

How much more for

 ? $ ________

5.  

$ ________ $ ________

How much more for

 ? $ ________

6.  

$ ________ $ ________

How much more for

 ? $ ________

SUBTRACTION

Use prices from the menu.
Solve each problem.

1.  

$ ________ $ ________

How much less for

 ? $ ________

4.  

$ ________ $ ________

How much less for

 ? $ ________

2.  

$ ________ $ ________

How much less for

 ? $ ________

5.  

$ ________ $ ________

How much less for

 ? $ ________

3.  

$ ________ $ ________

How much less for

 ? $ ________

6.  

$ ________ $ ________

How much less for

 ? $ ________

SUBTRACTION

Use prices from the menu.
Solve each problem.

1. Jan bought 

   Tim bought 

   How much more did Jan spend?

   $ ________

2. Lee bought 

   Rick bought 

   How much more did Lee spend?

   $ ________

3. Tom bought 

   Lou bought 

   How much more did Tom spend?

   $ ________

4. Ted bought 

   Dave bought 

   How much more did Dave spend?

   $ ________

5. Ray bought 

   Roy bought 

   How much more did Roy spend?

   $ ________

6. Jane bought 

   Mike bought 

   How much more did Mike spend?

   $ ________

SUBTRACTION

Use prices from the menu.
Solve each problem.

---

1. Jack had $2.25. He bought a  .

   How much money left? $ ____________

2. Tim had $1.16. He bought  .

   How much money left? $ ____________

3. Kate had $3.64. She bought a  .

   How much money left? $ ____________

4. Judy had $3.20. She bought a  .

   How much money left? $ ____________

5. Ramon had $1.66. He bought  .

   How much money left? $ ____________

6. Andy had $4.13. He bought  .

   How much money left? $ ____________

SUBTRACTION

Use prices from the menu.
Solve each problem.

1. You have $1.00

   You buy 

   What is your change?

   $ ______

2. You have $2.00

   You buy 

   What is your change?

   $ ______

3. You have $3.00

   You buy 

   What is your change?

   $ ______

4. You have $4.00

   You buy 

   What is your change?

   $ ______

5. Had $5.00

   You buy 

   What is your change?

   $ ______

6. You have $6.00

   You buy 

   What is your change?

   $ ______

7. You have $7.00

   You buy 

   What is your change?

   $ ______

8. You have $8.00

   You buy 

   What is your change?

   $ ______

SUBTRACTION

Use prices from the menu.
Solve each problem.

1. Pete bought pizza. Paul had shrimp.

   How much more did Paul spend? $ ________

2. John has $.94. He wants to buy a hot dog.

   How much more does he need? $ ________

3. Julie has spaghetti. Jane has chicken.

   How much more does chicken cost? $ ________

4. Jerry had $2.00. He bought an ice cream cone.

   How much money is left? $ ________

5. Marie has a hamburger. Her friend has onion

   rings. How much more does Marie pay? $ ________

SUBTRACTION

Use prices from the menu.
Solve each problem.

1. Alex bought a taco. Matt had a hamburger. How much more did Matt spend? $ __________

2. Eric had $3.15. He bought a banana split. How much money is left? $ __________

3. Dan wants a milk shake. He has $.28. How much more money does he need? $ __________

4. Sally used $5.00 to pay for her pancakes. How much change did she get? $ __________

5. Russ had a sundae. Ted had juice. How much more did Russ spend? $ __________

6. Fred wants spaghetti. He has $.79. How much more money does he need? $ __________

ADDITION - SUBTRACTION

Use prices from the menu.
Solve each problem.

1. You buy a 

   and a 

   How much in all?

   $ ____________

4. How much for

   this food? 

   $ ____________

2. How much more does

    cost

   than a  ?

   $ ____________

5. Had $2.25

   Bought 

   How much left?

   $ ____________

3. Dan bought 

   and 

   What is the total cost?

   $ ____________

6. Dave bought 

   Dick bought 

   How much more did
   Dave spend?

   $ ____________

ADDITION - SUBTRACTION

Use prices from the menu.
Solve each problem.

1. Jay bought a taco. Jan bought pizza.

   How much more did Jan spend? $ ________

2. How much for chili and onion rings? $ ________

3. Randy buys cake and hot chocolate.

   What is the total cost? $ ________

4. Stan used $3.00 to pay for a banana split.

   What is his change? $ ________

5. How much more does a hamburger cost than

   a hot dog? $ ________

6. How much for cake and a glass of milk? $ ________

ADDITION - SUBTRACTION
2 steps

Use prices from the menu.
Solve each problem.

Money to Start With	Food Bought	Money to Take Home
$1.64	 	
$3.83	 	
$3.74	 	
$2.82	 	
$5.91	 	
$7.59	 	
$6.50	 	

ADDITION - SUBTRACTION
2 steps

Use prices from the menu.
Solve each problem.

1. James had  and  .

   (a) How much did he spend? $ ____________

   (b) How much change did he get from $4.00? $ ____________

2. Bill ordered  and  .

   (a) How much did his food cost? $ ____________

   (b) How much change did he get from $4.00? $ ____________

3. Jean ate  and  .

   (a) How much did that cost? $ ____________

   (b) What was her change from $5.00? $ ____________

4. Amy had  and  .

   (a) How much did she spend? $ ____________

   (b) How much change from $5.00? $ ____________

ADDITION - SUBTRACTION
2 steps

Use prices from the menu.
Solve each problem.

1. Mike had $2.00. He bought a cola and french fries. How much money left? $ ________

2. Tony had $5.00. He ordered a hot dog and milk. How much change? $ ________

3. Kim had spaghetti and juice. She paid with $3.00. How much change? $ ________

4. Art had $6.25. He bought a sandwich and root beer. How much money left? $ ________

5. Mark ordered a hamburger and a milk shake. He paid with $3.75. How much change? $ ________

6. Steve used $5.00 to pay for 2 banana splits. What was his change? $ ________

MULTIPLICATION

Use prices from the menu.
Find the total cost.

1. Three  cost $ ________

2. Four  cost $ ________

3. Two  cost $ ________

4. Five  cost $ ________

5. Three  cost $ ________

6. Two  cost $ ________

7. Five  cost $ ________

8. Three  cost $ ________

9. Four  cost $ ________

10. Five  cost $ ________

MULTIPLICATION

Use prices from the menu.
Find the total cost.

1. Six  cost $ ________
2. Seven  cost $ ________
3. Eight  cost $ ________
4. Six  cost $ ________
5. Nine  cost $ ________
6. Seven  cost $ ________
7. Eight  cost $ ________
8. Nine  cost $ ________
9. Six  cost $ ________
10. Seven  cost $ ________

MULTIPLICATION

Use prices from the menu.
Find the total cost.

1. Three colas cost $ ____________
2. Four milk shakes cost $ ____________
3. Five lemonades cost $ ____________
4. Six hot dogs cost $ ____________
5. Two sandwiches cost $ ____________
6. Six sundaes cost $ ____________
7. Eight ice cream cones cost $ ____________
8. Nine tacos cost $ ____________
9. Five banana splits cost $ ____________
10. Seven hamburgers cost $ ____________

MULTIPLICATION

Use prices from the menu.
Find the total cost.

1. A man bought 4 hamburgers.
   How much did he spend? $ ____________

2. How much would 5 milk shakes cost? $ ____________

3. What would be the cost of 6 tacos? $ ____________

4. Joe bought 3 hot dogs.
   How much did he spend? $ ____________

5. A girl paid for 5 glasses of juice.
   How much did she spend? $ ____________

6. Find the total cost of 7 sandwiches. $ ____________

7. How much for 4 banana splits? $ ____________

ADDITION - SUBTRACTION - MULTIPLICATION

Use prices from the menu.
Solve each problem.

1. Buy  ,  , and  .
   How much in all? $ ________

2. How much more for  than for  ? $ ________

3. Five  cost how much? $ ________

4. Pay for  with $2.00.
   How much change? $ ________

5. Had $1.00. Bought an  .
   How much left? $ ________

6. How much for  ,  , and  ? $ ________

7. Six  cost how much? $ ________

ADDITION - SUBTRACTION - MULTIPLICATION

Use prices from the menu.
Solve each problem.

1. Find the total cost of    . $ ____________

2. How much for eight  ? $ ____________

3. You have $3.72. You buy  .
   How much left? $ ____________

4. A girl bought  and  .
   How much did she spend? $ ____________

5. You buy  . Your friend buys  .
   How much more is the  ? $ ____________

6. How much for nine  ? $ ____________

7. You want  . You have $.46.
   How much more do you need? $ ____________

ADDITION - SUBTRACTION - MULTIPLICATION

Use prices from the menu.
Solve each problem.

1. Gary had a sandwich and lemonade.

   How much did he spend? $ ________

2. Find the cost of six tacos. $ ________

3. How much more is bacon and eggs than

   pancakes? $ ________

4. What is the total cost of onion rings, a hot

   dog, and a banana split? $ ________

5. Dennis had $2.00. He bought a milk shake.

   How much money left? $ ________

6. How much for seven hamburgers? $ ________

7. Find the cost of a sundae and spaghetti. $ ________

ADDITION - SUBTRACTION - MULTIPLICATION

Use prices from the menu.
Solve each problem.

1. How much for seven sundaes? $ ____________

2. Find the cost of an ice cream cone, juice, and chili. $ ____________

3. You have $5.00. You buy a sandwich. How much is left? $ ____________

4. Your friend bought a hot dog and a milk shake. How much did she spend? $ ____________

5. A boy has $1.36. He wants to buy a hamburger. How much more does he need? $ ____________

6. How much more is a milk shake than an ice cream cone? $ ____________

ADDITION

Use prices from the menu.
Find the total cost.

Guest Check

1	Hamburger	
1	Root Beer	
	Total	

Guest Check

2	Hot Dogs	
2	Colas	
	Total	

ADDITION

Use prices from the menu.
Find the total cost.

Guest Check		
2	Sandwiches	
2	Milk Shakes	
	Total	

Guest Check		
2	Pizzas	
2	Milk	
	Total	

ADDITION

Use prices from the menu.
Find the total cost.

Guest Check

2	Ice Cream Cones	
3	Hot Dogs	
	Total	

Guest Check

2	French Fries	
3	Colas	
3	Sundaes	
	Total	

ADDITION

Use prices from the menu.
Find the total cost.

Guest Check

2	Banana Splits	
3	Sandwiches	
4	Hot Dogs	
	Total	

Create your own check.

Guest Check

	Total	

ADDITION - SUBTRACTION
2 steps

Use prices from the menu.
Find the total cost.

	Guest Check	
1	Hamburger	
1	Onion Rings	
1	Sundae	
	Total	

Pay with $6.00

Change __________

	Guest Check	
1	Pizza	
1	Lemonade	
1	Sandwich	
	Total	

Pay with $7.00

Change __________

Answer Key

P. 1
1) 1.75
2) 1.66
3) .92
4) 4.74
5) .90
6) .77
7) 4.10
8) .94
9) 1.63
10) 1.86
11) 1.69
12) 1.10
13) 2.16
14) 1.17
15) 3.99
16) 1.89

P. 2
1) 2.16
2) .74
3) .88
4) 1.63
5) 1.75
6) 4.43
7) 4.74
8) .92
9) 1.66
10) 4.10
11) 1.17
12) .75
13) .95
14) 1.10
15) .98
16) 1.52
17) 1.89
18) 1.69
19) 2.55
20) .77

P. 3
1) 2.16
+1.52
3.68

2) 1.75
+1.10
2.85

3) 1.10
+ .74
1.84

4) 4.10
+ .88
4.98

5) 4.43
+1.52
5.95

6) 2.16
+1.10
3.26

7) 4.10
+1.63
5.73

8) 4.74
+4.10
8.84

P. 4
1) 3.68
2) 6.29
3) 2.85
4) 5.00
5) 2.44
6) 5.68
7) 1.78
8) 5.84
9) 3.79
10) 1.94

P. 5
1) 4.32
2) 1.84
3) 1.76
4) 1.80
5) 1.96
6) 3.32
7) 2.34
8) 3.38
9) 1.48
10) 3.50

P. 6
1) 2.16
+1.66
3.82

2) 2.55
+ .75
3.30

3) 4.10
+4.43
8.53

4) 1.52
+1.63
3.15

5) 1.86
+1.69
3.55

6) 1.75
+ .74
2.49

7) .77
+4.74
5.51

8) 3.99
+ .95
4.94

P. 7
1) 2.79
2) 2.04
3) 3.38
4) 5.73
5) 3.08
6) 5.68
7) 6.06
8) 2.49
9) 6.40
10) 3.72

P. 8
1) 1.67
2) 4.84
3) 5.35
4) 1.98
5) 3.03
6) 6.37
7) 2.40

P. 9
1) 5.38
2) 3.30
3) 2.20
4) 1.64
5) 2.44
6) 2.64
7) 2.66

P. 10
1) .90
1.17
+ .98
3.05

2) 1.52
2.16
+ .88
4.56

3) 1.63
4.74
+1.69
8.06

4) 2.55
.94
+1.75
5.24

5) .92
4.10
+1.10
6.12

6) 4.43
.77
+3.99
9.19

P. 11
1) 10.28
2) 4.97
3) 4.19
4) 4.66
5) 6.68
6) 3.53
7) 9.91
8) 3.00

P. 12
1) 3.84
2) 7.14
3) 5.07
4) 5.92
5) 2.74
6) 7.31

P. 13
1) 3.39
2) 2.63
3) 6.87
4) 7.48
5) 8.26
6) 4.50

P. 14
1) .98
.92
+1.66
3.56

2) .77
1.86
+ .95
3.58

3) 4.43
.90
+1.52
6.85

4) 4.74
1.63
+ .74
7.11

5) 2.55
.75
+ .94
4.24

6) 1.10
1.17
+2.16
4.43

P. 15
1) 2.40
2) 4.24
3) 2.46
4) 4.51
5) 3.75
6) 4.93

P. 16
1) 4.25
2) 7.29
3) 3.23
4) 7.95
5) 7.25
6) 4.21

P. 17
1) 1.42
2) 1.34
3) 3.58
4) 3.05
5) 3.02
6) 4.22
7) 1.14
8) 1.65

P. 18
1) .11
2) .12
3) .12
4) .12
5) 1.04
6) 1.11
7) 1.21
8) 1.32

P. 19
.20
.63
1.09
1.07
1.43
.37
.24

P. 20
1) pizza .07
2) bacon & eggs 2.13
3) hamburger .09
4) root beer .02
5) onion rings .89

P. 21
1) 2.16
1.10
1.06

2) 1.75
.90
.85

3) 1.52
.94
.58

4) 1.89
.88
1.01

5) 4.74
4.10
.64

6) 1.69
4.43
2.74

P. 22
1) .90
4.10
3.20

2) 3.99
1.52
2.47

3) 1.69
4.74
3.05

4) .94
1.10
.16

5) 2.16
1.66
.50

6) 1.86
1.63
.23

P. 23
1) .77
2) .50
3) .62
4) .16
5) .40
6) 2.47

P. 24
1) .50
2) .26
3) 1.48
4) 2.10
5) .92
6) 2.44

P. 25
1) .23
2) .48
3) 1.31
4) 3.02
5) 1.01
6) 5.08
7) 2.57
8) 6.34

P. 26
1) 2.93
2) .16
3) 3.05
4) 1.12
5) .12

P. 27
1) .77
2) .99
3) 1.58
4) 2.45
5) .91
6) .90

P. 28
1) 6.09
2) 3.01
3) 5.27
4) 3.53
5) .09
6) .85

P. 29
1) .19
2) 3.52
3) 2.46
4) .84
5) .65
6) 2.47

P. 30
.13
2.01
.39
.31
1.89
1.41
2.98

P. 31
1a) 3.49
1b) .51
2a) 3.06
2b) .94
3a) 4.93
3b) .07
4a) 4.87
4b) .13

P. 32
1) .36
2) 2.95
3) .56
4) .90
5) .14
6) .68

P. 33
1) 6.48
2) 6.52
3) 3.38
4) 4.50
5) 11.97
6) 2.34
7) 7.60
8) 2.22
9) 7.56
10) 4.70

P. 34
1) 9.78
2) 5.18
3) 7.84
4) 5.52
5) 10.53
6) 6.16
7) 7.20
8) 14.94
9) 11.34
10) 5.39

P. 35
1) 2.22
2) 7.44
3) 3.85
4) 6.60
5) 8.86
6) 9.96
7) 7.04
8) 8.82
9) 10.80
10) 12.25

P. 36
1) 7.00
2) 9.30
3) 5.88
4) 3.30
5) 3.75
6) 31.01
7) 8.64

P. 37
1) 4.49
2) .50
3) 8.75
4) .14
5) .12
6) 7.09
7) 6.60

P. 38
1) 6.82
2) 5.92
3) 2.06
4) 2.55
5) .98
6) 13.68
7) .71

P. 39
1) 5.20
2) 5.88
3) 1.44
4) 4.89
5) .14
6) 12.25
7) 3.35

P. 40
1) 11.62
2) 3.52
3) .57
4) 2.96
5) .39
6) .98

P. 41
1.75
+ .92
2.67

2.20
+1.48
3.68

P. 42
8.86
+3.72
12.58

2.34
+1.90
4.24

P. 43
1.76
+3.30
5.06

1.80
2.22
+4.98
9.00

P. 44
4.32
13.29
+ 4.40
22.01

P. 45
1.75
1.63
+1.66
5.04

change .96

1.17
.77
+4.43
6.37

change .63